Patricia Kite *Down in the Sea:*

THE JELLYFISH

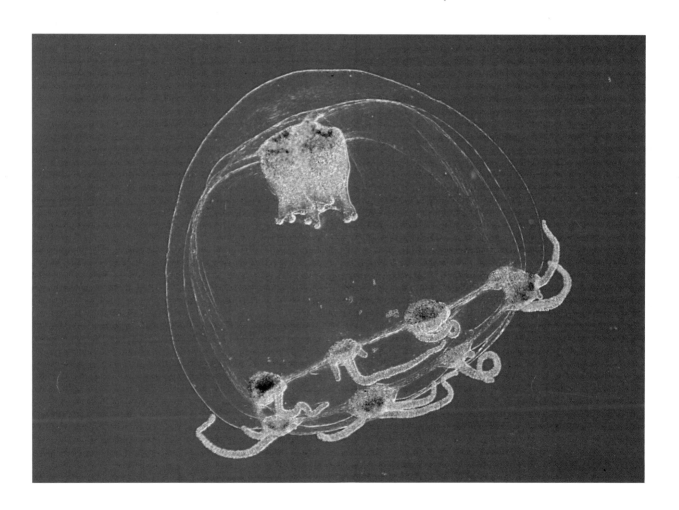

ALBERT WHITMAN & COMPANY • Morton Grove, Illinois

To my beloved uncles: Louis, Eli,
Ted, and Abe Evenchick

Thanks to Dustin Chivers, Senior Curatorial Assistant,
Department of Invertebrate Zoology,
California Academy of Sciences,
San Francisco, for his help.

Published in 1993
by Albert Whitman & Company,
6340 Oakton, Morton Grove, IL 60053-2723.
Published simultaneously in Canada
by General Publishing, Limited, Toronto.

Library of Congress Cataloging-in-Publication
Data
Kite, L. Patricia. 1940–
Down in the sea: The jellyfish / L. Patricia Kite.
p.   cm.
Summary: A simple introduction to the jellyfish,
describing its physical characteristics, life cycle,
and eating habits.
ISBN 0-8075-1712-7
1. Jellyfishes—Juvenile literature. [1. Jellyfishes.]
I. Title. II. Title: Jellyfish.
QL377.S4K62 1993   92-12834
593.7—dc20   CIP   AC

Cover and interior design:
Karen A. Yops.

The text typeface is Optima.

The cover shows a jellyfish
cousin, the Porpita.

A jiggly jellyfish
is flopped on the beach.
Do not poke it. Ouch!

Soon the sun
dries out the jelly,
which is really mostly water.
The next day,
if you looked,
all you would see is this.

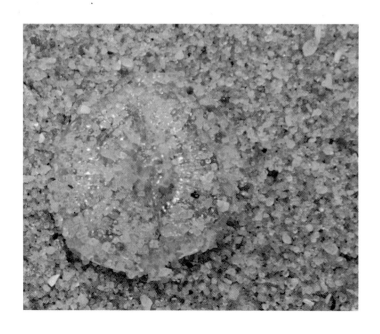

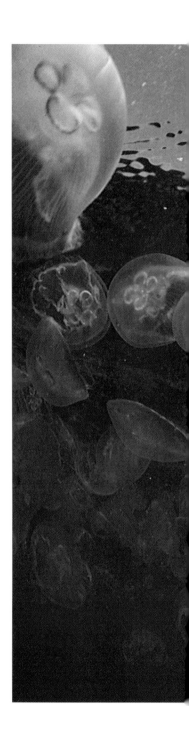

But there are
many more jellyfish
floating in the water,
near, or far, far away.

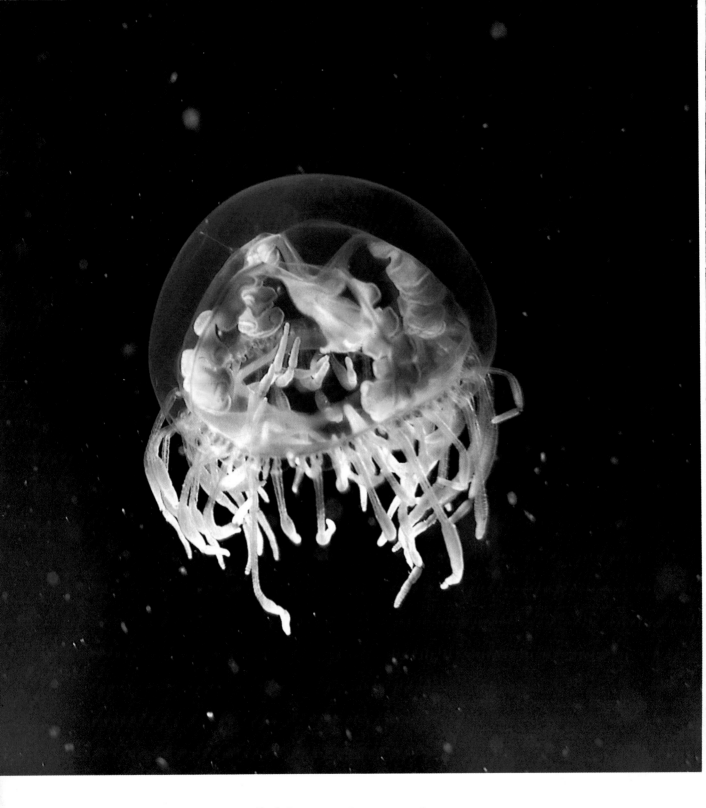

Jellyfish come in many sizes—
smaller than a grape
and bigger than a bed.

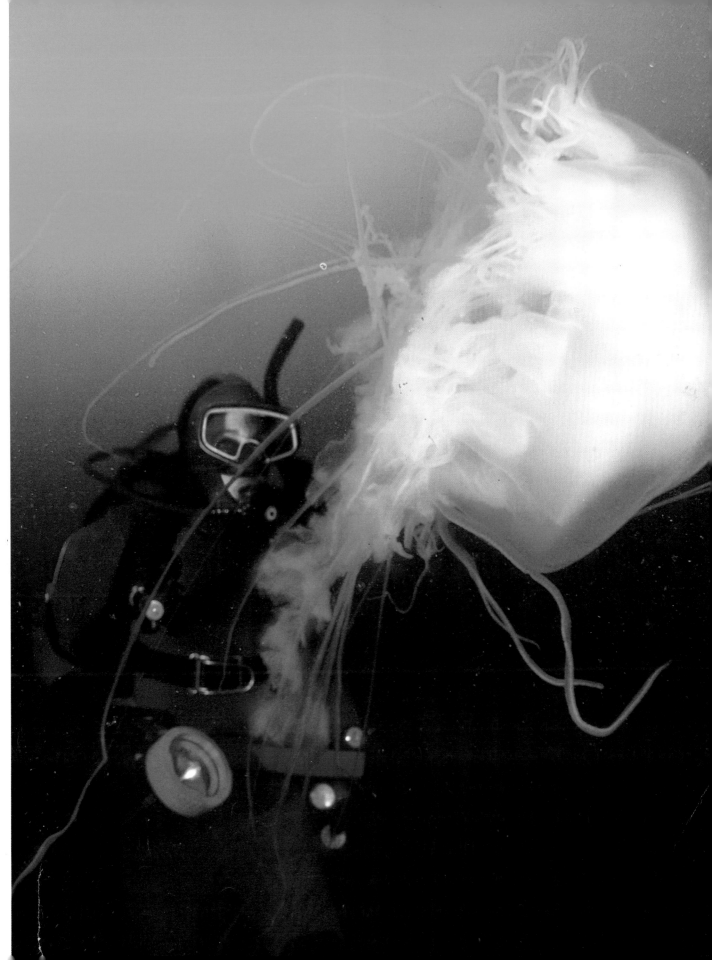

Jellyfish come in many shapes.
Some look like cups.
Some look like bowls.
Some look like parachutes.

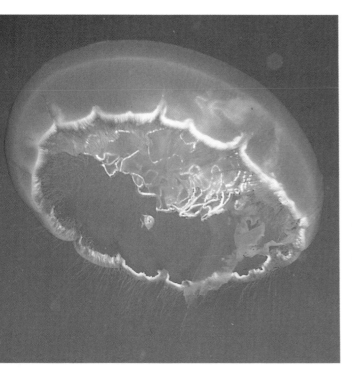

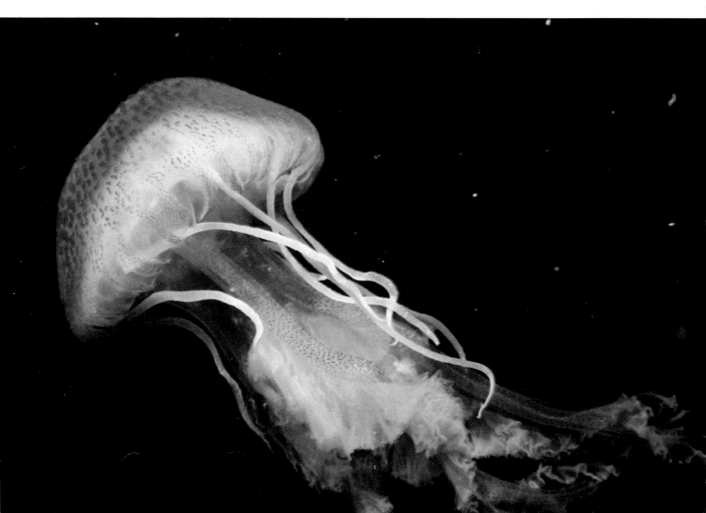

Many jellyfish are see-through,
but they can come in colors—
white, blue, pink, yellow, green,
purple, red, orange, and striped.

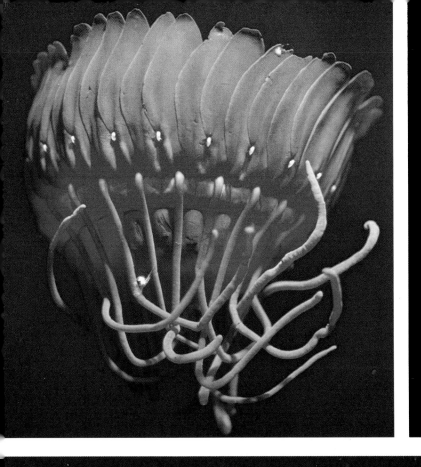

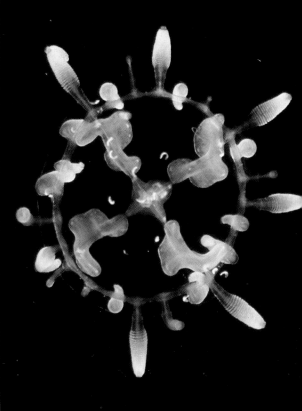

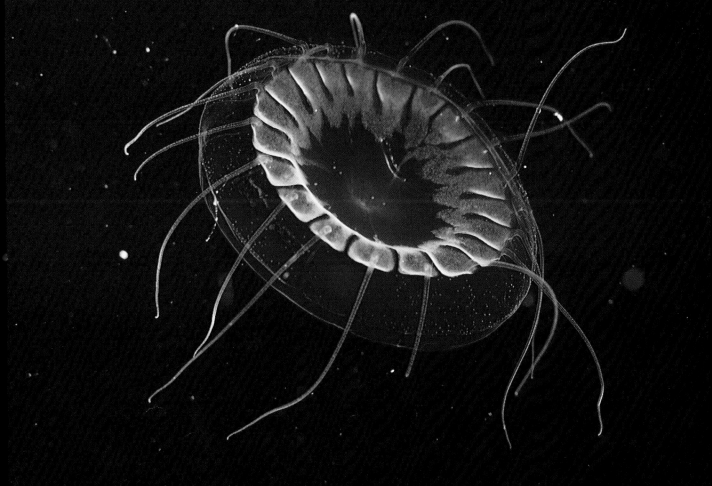

Some even light up at night
or in sunless deep water.

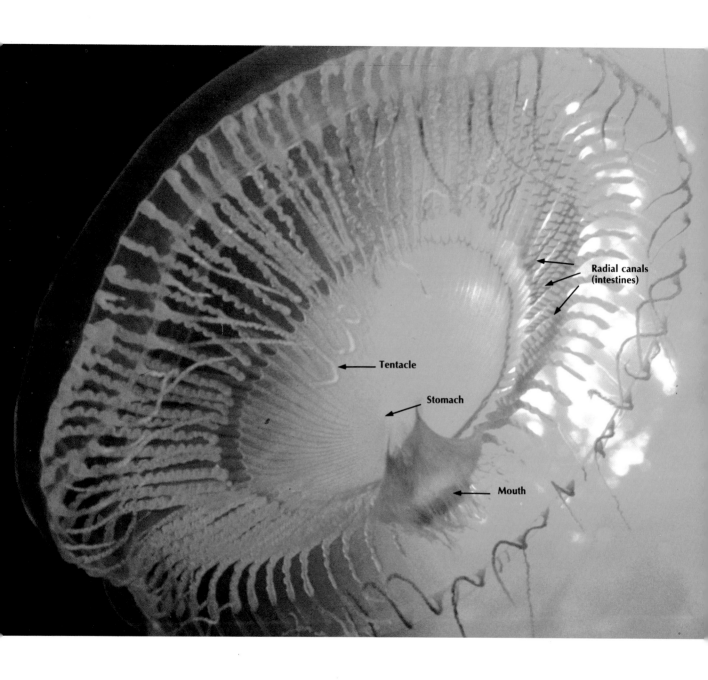

Radial canals
(intestines)

Tentacle

Stomach

Mouth

Jiggly jellyfish
look like blobs of jelly
around a stomach hole.
All eat a lot.

Jellyfish eat fish, crabs, worms,
shrimp, plankton, plants,
and, sometimes, smaller jellyfish. Yum!

Most jellyfish catch food with tentacles.
Tentacles look like cooked noodles
hanging from under a jellyfish jiggly body.

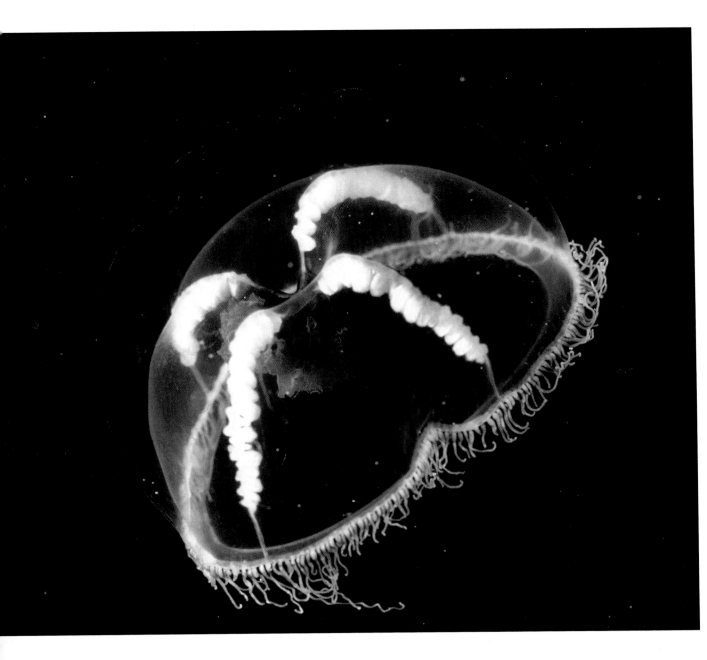

Some tentacles are just an inch or less.
But they can be one hundred twenty feet long.
That's longer than a basketball court. Wow!

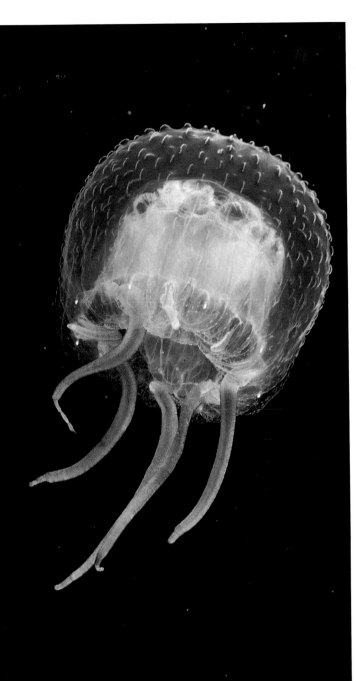
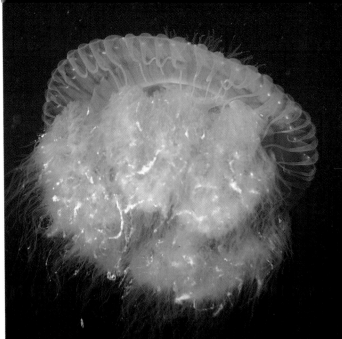
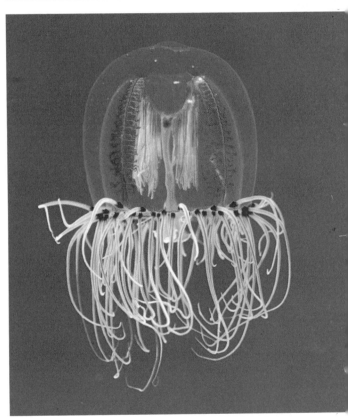

A jellyfish may have a few tentacles
or as many as eight hundred.

Each tentacle is dotted
with stinging cells.
When food swims by,
it touches a tentacle.
Ouch! Or even ouch! ouch! ouch!
Soon the food no longer moves.
The tentacles pull it
to the jellyfish mouth.

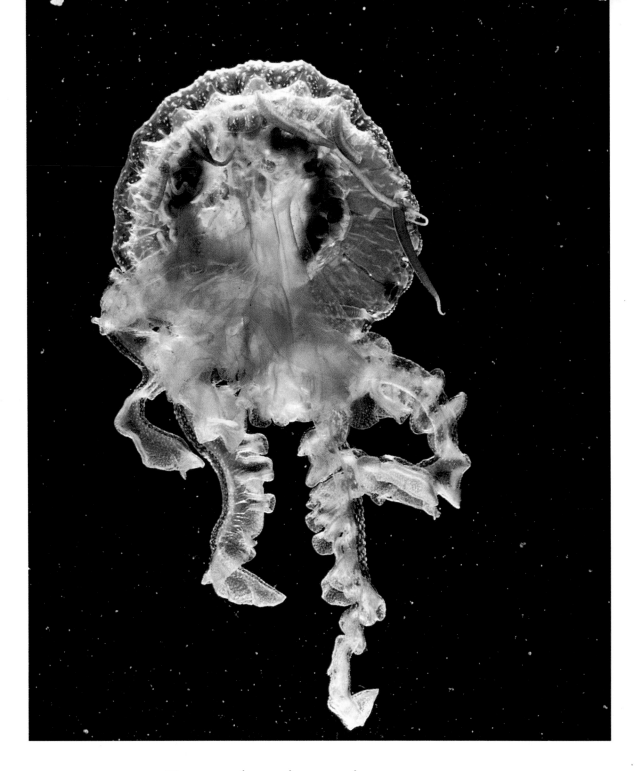

To move from place to place,
a jellyfish floats or opens
and closes its body like an umbrella.
Opening lets water under the umbrella.
Closing pushes it out like a jet. Whoosh!

Baby jellyfish don't look
like their parents.
At least, not at first.
This is how many jellyfish grow.

Each jellyfish lives about a year,
if a fish or turtle or bird
does not eat it first,
and the water is not too warm or cold.
But jellyfish have been
on this earth a long time
(four hundred fifty million years).
There were jellyfish
before there were dinosaurs!

And some of them melted
on a beach, too,
until they looked like this.

ABOUT JELLYFISH

Tiny, medium, big—and huge. There are thousands of different jellyfish types. They live throughout the world in both fresh and salt water, in oceans, bays, lakes, swamps, and even ponds.

Jellyfish are *invertebrates,* or creatures without backbones. They do not have a central brain like ours. But they can relay body-movement messages within their simple-looking nervous systems. Instead of just drifting, many jellyfish may move deliberately in search of food.

Special cells within the jellyfish body detect food odors. Eyespots can sense the difference between light and dark. The *Cubomedusae* jellyfish have advanced eye forms that actually perceive images.

All jellyfish have stinging cells on their tentacles. At rest, these cells resemble coiled, threadlike tubes, each contained within a separate capsule. Some tubes have barbed edges. The coiled threads can spring out faster than anyone can see. They tear into prey and give out a poisonous chemical. A few jellyfish have chemicals in their stinging cells that can harm or even kill people.

Many jellyfish also shoot out nonpoisonous threads. These wrap around prey, holding it in place until it is brought to the mouth hole.

Jellyfish body "jelly" is contained within a pouch or bag. The jelly is about 95 percent water and 4 percent salts. The other 1 percent is made up of many different chemicals.

The jelly pouch has a thin, protective outside layer. The inside layer is made of digestive cells. Some of these cells give off chemicals that help dissolve food into very tiny pieces. Water entering through the mouth hole moves this food through the jellyfish body. Waste is pushed back out through the mouth hole.

Since jellyfish are not really fish, many scientists today prefer you use the term *jellies* instead. But whatever you call these creatures, don't poke or step on any washed up on the beach. Jellyfish that look dead can still have stinging cells that work.

When jellyfish die, they leave behind no hard parts, such as shells or bones, for a fossil historical record. But a very long time ago, some died in mud or sand that dried hard. Some of the resulting body impressions date back four hundred fifty million years.

CREDITS: Cover and p. 20: Animals Animals/© Kathie Atkinson, OSF; p. 1: © Paulette Brunner/Tom Stack & Associates; p. 2: © Tom McHugh/Photo Researchers, Inc.; pp. 3, 4 (above and below), 9 (bottom), 13, 14, 22: © Herb Segars; pp. 4–5: © Fred Bavendam; pp. 6, 17: © F. Stuart Westmorland; pp. 7, 9 (top right), 18 (left): © Dave B. Fleetham/Tom Stack & Associates; pp. 8, 11 (bottom): © Neil McDaniel; p. 9 (top left): © John Lidington/Photo Researchers, Inc.; pp. 10, 11 (top left), 15: © William Curtsinger/Photo Researchers, Inc.; pp. 11 (top right), 16, 18 (bottom right): © Sea Studios; p. 12: © David Hall; p. 18 (top right): © Carl Roessler; p. 19: Animals Animals/© Peter Parks, OSF; p. 21: art © Valerie A. Kells; p. 23: Animals Animals/ © Fred Whitehead.